YOUR KNOWLEDGE HAS VALUE

- We will publish your bachelor's and
  master's thesis, essays and papers

- Your own eBook and book -
  sold worldwide in all relevant shops

- Earn money with each sale

Upload your text at www.GRIN.com
and publish for free

Kishan Panaganti

Audio source separation using independent component analysis and beam formation

GRIN Verlag

Bibliografische Information der Deutschen Nationalbibliothek:

Die Deutsche Bibliothek verzeichnet diese Publikation in der Deutschen National-
bibliografie; detaillierte bibliografische Daten sind im Internet über http://dnb.d-
nb.de/ abrufbar.

Imprint:

Copyright © 2013 GRIN Verlag GmbH
Druck und Bindung: Books on Demand GmbH, Norderstedt Germany
ISBN: 978-3-656-58886-3

AUDIO SOURCE SEPERATION

USING INDEPENDENT

COMPONENT ANALYSIS AND

ACOUSTIC BEAM FORMATION

Kishan P B

Dept. of ECE

2014

P.E.S Institute of Technology
Bengaluru-85

TABLEOFCONTENTS:

DEPT OF ECE

DEPT OF ECE

ABSTRACT:

Audio source separation is the problem of automated separation of audio sources present in a room, using a set of differently placed microphones, capturing the auditory scene. The whole problem resembles the task a human can solve in a cocktail party situation, where using two sensors (ears), the brain can focus on a specific source of interest, suppressing all other sources present (cocktail party problem).

For computational and conceptual simplicity this problem is often represented as a linear transformation of the original audio signals. In other words, each component (multivariate signal) of the representation is a linear combination of the original variables (original subcomponents).

In signal processing, independent component analysis (ICA) is a computational method for separating a multivariate signal into additive subcomponents by assuming that the subcomponents are non-Gaussian signals and that they are all statistically independent from each other. Such a representation seems to capture the essential structure of the data in many applications.

Here we separate audio using different criteria suggested for ICA, being PCA (Principal Component Analysis), Non-gaussianity maximization using kurtosis and neg-entropy methods, frequency domain approach using non-gaussianity maximization and beamforming.

LIST OF FIGURES:

DEPT OF ECE

1. <u>MOTIVATION</u>

Imagine that you are in a room where two people are speaking simultaneously. You have two microphones, which you hold in different locations. The microphones give you two recorded time signals, which we could denote by $x_1(t)$ and $x_2(t)$, with x_1 and x_2 the amplitudes, and t the time index. Each of these recorded signals is a weighted sum of the speech signals emitted by the two speakers, which we denote by $s_1(t)$ and $s_2(t)$. We could express this as a linear equation:

$$x_1(t) = a_{11}s_1 + a_{12}s_2 \dots\dots (1)$$
$$x_2(t) = a_{21}s_1 + a_{22}s_2 \dots\dots (2)$$

where a_{11}, a_{12}, a_{21}, and a_{22} are some parameters that depend on the distances of the microphones from the speakers. It would be very useful if you could now estimate the two original speech signals $s_1(t)$ and $s_2(t)$, using only the recorded signals $x_1(t)$ and $x_2(t)$. This is called the cocktail-party problem.

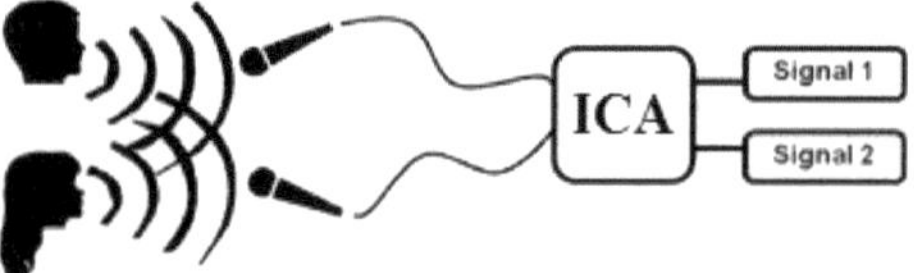

Figure 1: Signal mixing.

Here parameters a_{ii} and the original source signals s_i are both unknown. If by any chance parameters a_i are known signals s_i can be easily solved as solutions of linear equations. But a_{ii} are unknown and problem would be solved if we use some information on the statistical properties of the signals $s_i(t)$ to estimate the a_{ii}. It turns out that it is enough to assume that $s_1(t)$ and $s_2(t)$, at each time instant t, are statistically independent and non-gaussian. This is not an unrealistic assumption in many cases, and it need not be exactly true in practice. The recently developed technique of Independent Component Analysis (ICA), can be used to

DEPT OF ECE

estimate the a_{ij} based on the information of their independence, which allows us to separate the two original source signals $s_1(t)$ and $s_2(t)$ from their mixtures $x_1(t)$ and $x_2(t)$.

2. <u>INDEPENDENT COMPONENT ANALYSIS.</u>

<u>2.1 Definition of ICA</u> (ICA model)

Assume that there are n linear mixtures $x_1, ...,x_n$ of n independent components

$$x_j = a_{j1}s_1 + a_{j2}s_2 + ... + a_{jn}s_n, \text{ for all } j \ldots\ldots\ldots (3)$$

Dropping the time index t in the ICA model, assume that each mixture x_j as well as each independent component s_k is a random variable, instead of a proper time signal. The observed values $x_j(t)$, e.g., the microphone signals in the cocktail party problem, are then a sample of this random variable. Without loss of generality, we can assume that both the mixture variables and the independent components have zero mean. If this is not true, then the observable variables x_i can always be centered by subtracting the sample mean, which makes the model zero-mean.

It is convenient to use vector-matrix notation instead of the sums like in the previous equation. Let us denote by $\mathbf{x}$ the random vector whose elements are the mixtures $x_1, ...,x_n$, and likewise by $\mathbf{s}$ the random vector with elements $s_1, ..., s_n$. Let us denote by $\mathbf{A}$ the matrix with elements a_{ij}. Bold lower case letters indicate vectors and bold upper-case letters denote matrices. All vectors are understood as column vectors: thus $\mathbf{x}^T$, or the transpose of $\mathbf{x}$, is a row vector. Using this vector-matrix notation, the above mixing model is written as

$$\mathbf{x} = \mathbf{As} \ldots\ldots (4)$$

The statistical model in Eq. 4 is called independent component analysis, or ICA model. The ICA model is a generative model, which means that it describes how the observed data are generated by a process of mixing the components s_i. The independent components are latent variables, meaning that they cannot be directly observed. Also the mixing matrix is assumed

DEPT OF ECE

to be unknown. All we observe is the random vector $\mathbf{x}$, and we must estimate both $\mathbf{A}$ and $\mathbf{s}$ using it. This must be done under as general assumptions as possible.

2.2 Estimation approach

The starting point for ICA is the very simple assumption that the components s_i are statistically independent. Another assumption is that the independent component must have non-gaussian distributions. For simplicity, let the unknown mixing matrix is square, but this assumption can be sometimes relaxed. In many applications, it would be more realistic to assume that there is some noise in the measurements, which would mean adding a noise term in the model. For simplicity, we omit any noise terms and it seems to be sufficient for many applications.

Then, after estimating the matrix $\mathbf{A}$, we can compute its inverse, say $\mathbf{W}$, and obtain the independent component simply by En5 or $\mathbf{W}$ can be directly found without computing $\mathbf{A}$.

$$\mathbf{s} = \mathbf{W}\mathbf{x} \ \dots\dots.(5)$$

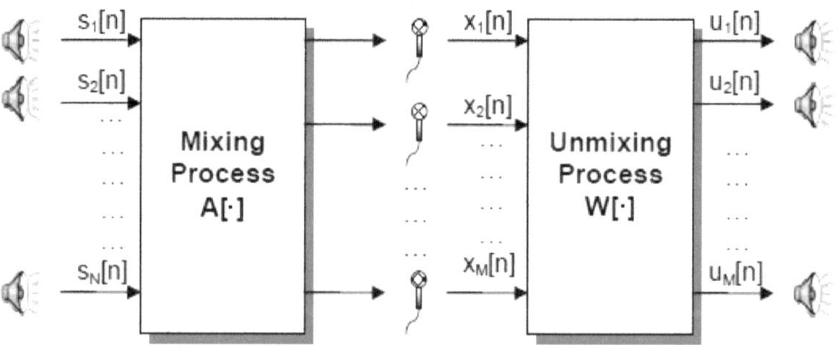

Figure 2: The general noiseless audio source separation problem.

Figure 2 pictorially represents the ICA model (ignoring noise) for audio source separation problem showing the unmixing process. The system outputs u_i are the estimated versions of s_i.

DEPT OF ECE

2.3 Ambiguities of ICA

In the ICA model in Eq. (4), it is easy to see that the following ambiguities will hold:

1. **Determining the variances (energies) of the independent components is difficult.**

The reason is that, both **s** and **A** being unknown, any scalar multiplier in one of the sources s_i could always be cancelled by dividing the corresponding column $\mathbf{a}_i$ of **A** by the same scalar. As a consequence, we fix the magnitudes of the independent components; as they are random variables, the most natural way to do this is to assume that each has unit variance: $E\{s_i^2\}= 1$. Then the matrix **A** will be adapted in the ICA solution methods to take into account this restriction. This still leaves the ambiguity of the sign: we could multiply the independent component by -1 without affecting the model. This ambiguity is, fortunately, insignificant in most applications.

2. **We cannot determine the order of the independent components.**

The reason is that, both **s** and **A** being unknown, we the order of the terms in the sum can be freely changed and call any of the independent components the first one. Formally, a permutation matrix **P** and its inverse can be substituted in the model to give $\mathbf{x} = \mathbf{AP}^{-1}\mathbf{Ps}$. The elements of **Ps** are the original independent variables s_j but in another order. The matrix $\mathbf{AP}^{-1}$ is just a new unknown mixing matrix, to be solved by the ICA algorithms.

But, these two ambiguities are insignificant for most applications thus can be ignored.

2.4 Illustration

Signals in Figure 5 are the estimates of signals from Figure 3 with only knowing signals in Figure 4. Here it can be noted that signal estimated can be inverted and scaled version of input signals.

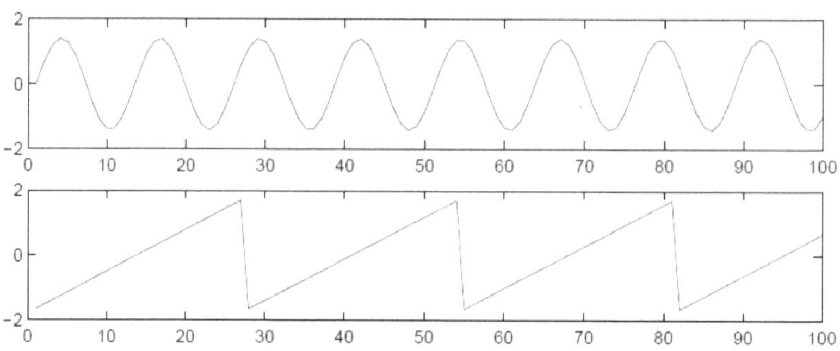

Figure 3: Original unmixed signals.

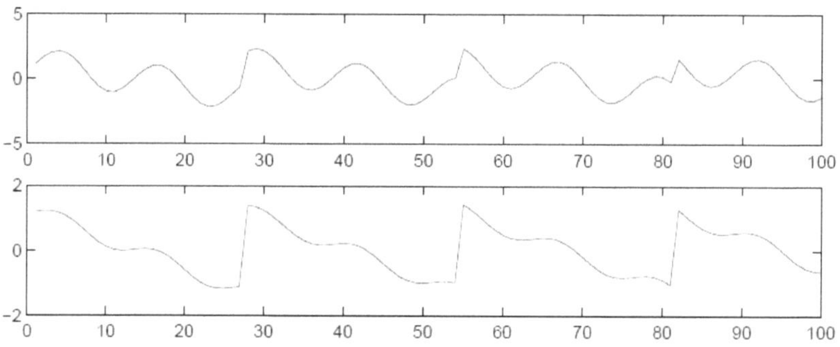

Figure 4: Mixed signals of original signals.

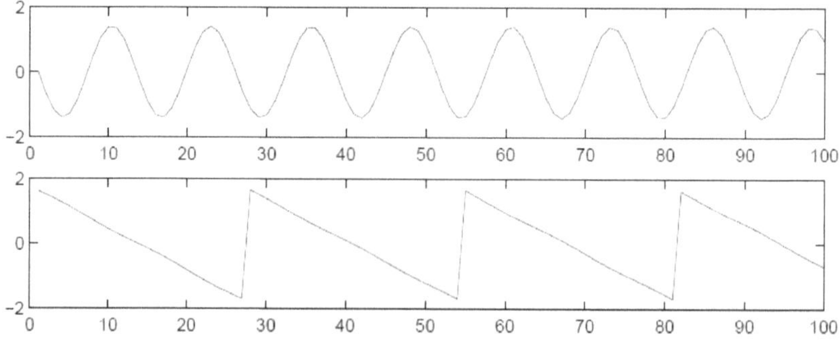

Figure 5: The estimates of the original source signals.

3. <u>Assumptions</u>

1. <u>Independence of source signals.</u>

By the very nature of the problem defined where audio of speakers is mixed the
sources are independent. That is, what each speaker speaks is independent of
other. Thus this assumption is practical.

Also signals which are independent are always uncorrelated while the other way
round is not true. The s_i's are independent while x_i's (linear combinations of s_i')
are dependent. If we decorrelate these x_i's we still do not get independent outputs
as by above property. Thus outputs are not great estimates of s_i's. This is called
PCA(Principal component Analysis) method of estimation. Below is an example
for two signal sources assumed to have uniform distribution.

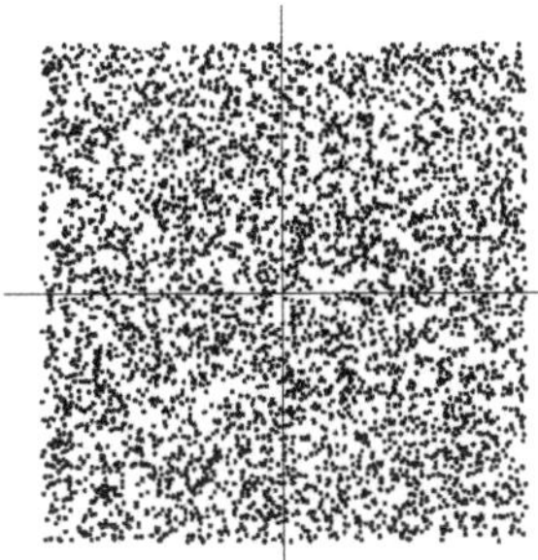

Figure 6: The joint distribution of
the independent components s_1
and s_2 with uniform distributions.
Horizontal
axis: s_1, vertical axis: s_2.

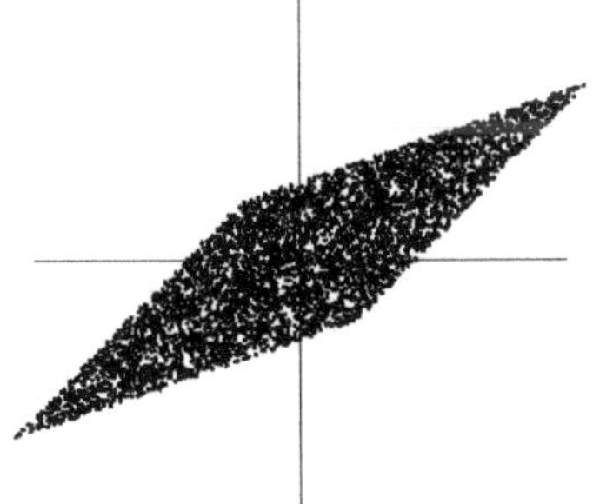

Figure 7: The joint distribution of
the observed mixtures x_1 and x_2.
Horizontal axis: x_1, vertical axis:
x_2.

DEPT OF ECE

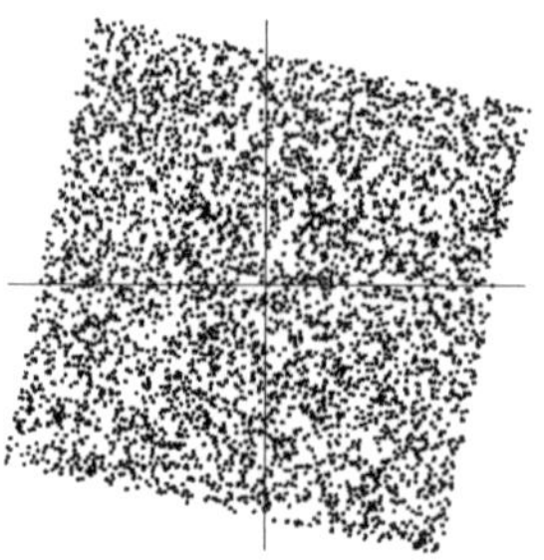

Figure 8: The joint distribution of the whitened mixtures.

From above figures it can be seen that Figure 8 is not same as Figure 6 and the signals of figure 8 are are not independent but are un correlated. Thus just **PCA** Is not a good estimation of signals. This leads towards the idea of *non-gaussianity maximization.*

2. Gaussian variables are forbidden

If the original signals are all Gaussian then the observed signal is also Gaussian as s_i are independent and observed signals are linear combinations of these signals. Since we use the concept of non-gaussianity maximization of observed signals are Gaussian it does not serve the purpose. Thus Gaussian signals (signals with Gaussian distribution) are forbidden.

4. ICA estimation using non-gaussianity maximization.

4.1 Principle

"Nongaussian is independent"

The Central Limit Theorem, a classical result in probability theory, tells that the distribution of a sum of independent random variables tends toward a gaussian distribution, under certain conditions. Thus, a sum of many independent random variables usually has a distribution that is closer to gaussian than any of the original random variables.

Assume that the data vector **x** is distributed according to the ICA data model in Eq. 4, i.e. it is a mixture of independent components. For simplicity, assumption is that all the independent components have identical distributions. To estimate one of the independent

DEPT OF ECE

components, we consider a linear combination of the x_i (eq. 6); let us denote this by $y = \mathbf{w}^T\mathbf{x}$ where $\mathbf{w}$ is a vector to be determined. If $\mathbf{w}$ were one of the rows of the inverse of $\mathbf{A}$, this linear combination would actually equal one of the independent components. Using CLT we have to find out a $\mathbf{w}$ such that it is one row of a^{-1}.

Defining a new transformation $\mathbf{z} = \mathbf{A}^T\mathbf{w}$ we have $y = \mathbf{w}^T\mathbf{x} = \mathbf{w}^T\mathbf{As} = \mathbf{z}^T\mathbf{s}$. y is thus a linear combination of s_i, with weights given by z_i. Since a sum of even two independent random variables is more gaussian than the original variables, $\mathbf{z}^T\mathbf{s}$ is more gaussian than any of the s_i and becomes least gaussian when it in fact equals one of the s_i. In this case, obviously only one of the elements z_i of $\mathbf{z}$ is nonzero. (Note that the s_i were here assumed to have identical distributions.) Therefore, we could take as $\mathbf{w}$ a vector that *maximizes the nongaussianity* of $\mathbf{w}^T\mathbf{x}$. Such a vector would necessarily correspond (in the transformed coordinate system) to a $\mathbf{z}$ which has only one nonzero component. This means that $\mathbf{w}^T\mathbf{x} = \mathbf{z}^T\mathbf{s}$ equals one of the independent components.

Maximizing the nongaussianity of $\mathbf{w}^T\mathbf{x}$ thus gives us one of the independent components. In fact, the optimization landscape for nongaussianity in the n-dimensional space of vectors $\mathbf{w}$ has $2n$ local maxima, two for each independent component, corresponding to s_i and $-s_i$ (recall that the independent components can be estimated only up to a multiplicative sign). To find several independent components, we need to find all these local maxima. This is not difficult, because the different independent components are uncorrelated: We can always constrain the search to the space that gives estimates uncorrelated with the previous ones. This corresponds to orthogonalization in a suitably transformed (i.e. whitened) space.

To use nongaussianity in ICA estimation, we must have a quantitative measure of non- gaussianity of a random variable. Some such measures are as discussed below.

4.2 Kurtosis as a measure of non-gaussianity

The classical measure of nongaussianity is kurtosis or the fourth-order cumulant. The kurtosis of y is classically defined by (where y is some random variable with zero mean and unit variance).

$$\text{kurt}(y) = E\{y^4\} - 3(E\{y^2\})^2 \quad \ldots\ldots (6)$$

DEPT OF ECE

Actually, since we assumed that y is of unit variance, the right-hand side simplifies to $E\{y^4\} - 3$.

This shows that kurtosis is simply a normalized version of the fourth moment $E\{y^4\}$. For a gaussian y, the fourth moment equals $3(E\{y^2\})^2$. Thus, kurtosis is zero for a gaussian random variable.

For most (but not quite all) nongaussian random variables, kurtosis is nonzero. Kurtosis can be both positive and negative. Random variables that have a negative kurtosis are called subgaussian, and those with positive kurtosis are called supergaussian. In statistical literature, the corresponding expressions platykurtic and leptokurtic are also used. Supergaussian random variables have typically a "spiky" pdf with heavy tails, i.e. the pdf is relatively large at zero and at large values of the variable, while being small for intermediate values. Subgaussian random variables, on the other hand, have typically a "flat" pdf, which is rather constant near zero, and very small for larger values of the variable.

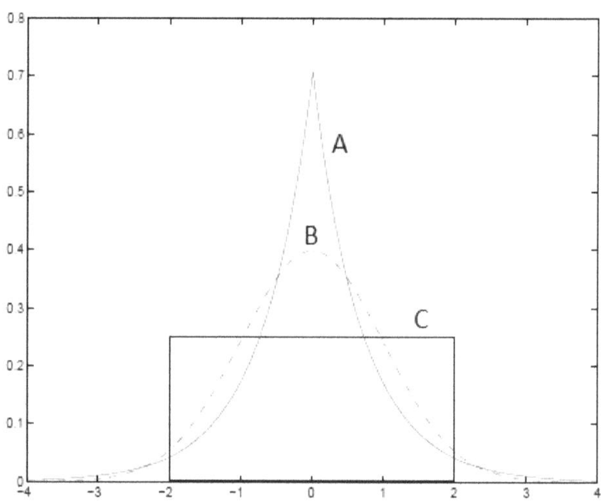

Typically nongaussianity is measured by the absolute value of kurtosis. The square of kurtosis can also be used. These are zero for a gaussian variable, and greater than zero for most nongaussian random variables. There are nongaussian random variables that have zero

DEPT OF ECE

kurtosis, but they can be considered as very rare. Kurtosis, or rather its absolute value, has been widely used as a measure of nongaussianity in ICA and related fields. The main reason is its simplicity, both computational and theoretical. Computationally, kurtosis can be estimated simply by using the fourth moment of the sample data.

In practice we would start from some weight vector $\mathbf{w}$, compute the direction in which the kurtosis of $y = \mathbf{w}^T \mathbf{x}$ is growing most strongly (if kurtosis is positive) or decreasing most strongly (if kurtosis is negative) based on the available sample $\mathbf{x}(1)$, ...,$\mathbf{x}(T)$ of mixture vector $\mathbf{x}$, and use a gradient method or one of their extensions for finding a new vector $\mathbf{w}$. The example can be generalized to arbitrary dimensions, showing that kurtosis can theoretically be used as an optimization criterion for the ICA problem. However, kurtosis has also some drawbacks in practice, when its value has to be estimated from a measured sample. The main problem is that kurtosis can be very sensitive to outliers. Its value may depend on only a few observations in the tails of the distribution, which may be erroneous or irrelevant observations. In other words, kurtosis is not a robust measure of nongaussianity.

Thus, other measures of nongaussianity might be better than kurtosis in some situations. Considering negentropy whose properties are rather opposite to those of kurtosis as a measure of non-gaussianity we proceed, and finally introduce approximations of negentropy that more or less combine the good properties of both measures.

4.3 Neg-Entropy as a measure of non-gaussianity

A second very important measure of nongaussianity is given by negentropy. It is based on the information theoretic quantity of (differential) entropy. Entropy is the basic concept of information theory. The entropy of a random variable can be interpreted as the degree of information that the observation of the variable gives. The more "random", i.e. unpredictable and unstructured the variable is, the larger its entropy.

Entropy H is defined for a discrete random variable Y a

$$H(Y) = - \sum P(Y = a_i) \log(P(Y = a_i)) \quad \ldots\ldots(7)$$

where the *ai* are the possible values of *Y*. This very well-known definition can be generalized for continuous-valued random variables and vectors, in which case it is often called differential entropy. The differential entropy *H* of a random vector **y** with density $f(\mathbf{y})$ is defined as

$$H(y) = - \int f(y)\log(f(y)) \, dy \ \ldots\ldots.(8)$$

A fundamental result of information theory is that *a gaussian variable has the largest entropy among all random variables of equal variance.* This means that entropy could be used as a measure of nongaussianity. In fact, this shows that the gaussian distribution is the "most random" or the least structured of all distributions. Entropy is small for distributions that are clearly concentrated on certain values, i.e., when the variable is clearly clustered, or has a pdf that is very "spiky". To obtain a measure of nongaussianity that is zero for a gaussian variable and always nonnegative, one often uses a slightly modified version of the definition of differential entropy, called negentropy. Negentropy *J* is defined as follows

$$J(\mathbf{y}) = H(\mathbf{y}_{gauss}) - H(\mathbf{y}) \ \ldots\ldots.(9)$$

where $\mathbf{y}_{gauss}$ is a Gaussian random variable of the same covariance matrix as **y**. Due to the above-mentioned properties, negentropy is always non-negative, and it is zero if and only if **y** has a Gaussian distribution.

The advantage of using negentropy, or, equivalently, differential entropy, as a measure of nongaussianity is that it is well justified by statistical theory. In fact, negentropy is in some sense the optimal estimator of nongaussianity, as far as statistical properties are concerned. The problem in using negentropy is, however, that it is computationally very difficult. Estimating negentropy using the definition would require an estimate (possibly nonparametric) of the pdf. Therefore, simpler approximations of negentropy are very useful, as will be discussed next.

4.3.1 Approximations of Negentropy

The estimation of negentropy is difficult, as mentioned above, and therefore this contrast function remains mainly a theoretical one. In practice, some approximation has to be used. Here we introduce approximations that have very promising properties, and which will be used in the following to derive an efficient method for ICA. The classical method of approximating negentropy is using higher-order moments, for example as follows

$$J(y) = (1/12)*E\{y^3\}^2 + (1/48)\, \mathrm{kurt}(y)^2 \quad \ldots\ldots(10)$$

The random variable y is assumed to be of zero mean and unit variance. However, the validity of such approximations may be rather limited. In particular, these approximations suffer from the nonrobustness encountered with kurtosis. To avoid the problems encountered with the preceding approximations of negentropy, new approximations were developed. These approximations were based on the maximum-entropy principle. In general we obtain the following approximation:

$$J(y) = \sum ki[E\{Gi(y)\} - E\{Gi(n)\}]^2, \quad \ldots\ldots\ldots(11)$$

where ki are some positive constants, and n is a Gaussian variable of zero mean and unit variance (i.e., standardized). The variable y is assumed to be of zero mean and unit variance, and the functions Gi are some nonquadratic functions . Note that even in cases where this approximation is not very accurate, (11) can be used to construct a measure of nongaussianity that is consistent in the sense that it is always non-negative, and equal to zero if y has a Gaussian distribution.

In the case where we use only one nonquadratic function G, the approximation becomes

$$J(y) = k*[E\{G(y)\} - E\{G(n)\}]^2 \quad \ldots\ldots\ldots\ldots(12)$$

for practically any non-quadratic function G. This is clearly a generalization of the moment-based approximation in(10) , if y is symmetric. Indeed, taking $G(y)=y4$, one then obtains exactly (10), i.e. a kurtosis-based approximation.

But the point here is that by choosing G wisely, one obtains approximations of negentropy that are much better than the one given by (10). In particular, choosing G that does not grow too fast, one obtains more robust estimators. The following choices of G have proved very useful:

$$G1(u) = (1/a_1)*\mathrm{logcosh}(a_1 u), \quad G2(u) = -\exp(-u^2/2) \ \ldots\ldots\ldots\ldots\ldots..(13)$$

Where, $1 \leq a1 < 2$ is some suitable constant. Thus we obtain approximations of negentropy that give a very good compromise between the properties of the two classical nongaussianity measures given by kurtosis and negentropy. They are conceptually simple, fast to compute, yet have appealing statistical properties, especially robustness.

A practical algorithm based on these contrast function and can be used for ICA. Any of the algorithm like gradient descent can be used for maximization. Here we use a fixed-point iteration scheme for finding a maximum of the nongaussianity of $\mathbf{w}^T\mathbf{x}$. Denote by g the derivative of the nonquadratic function G used in eq(12); for example the derivatives of the functions in eq(13) are:

$g1(u) = \tanh(a1u),$

$g2(u) = u\exp(-u2/2)$

where $1 \geq a1 \leq 2$ is some suitable constant, often taken as $a1 = 1$.

The basic form of the ICA algorithm is as follows:
1. Choose an initial (e.g. random) weight vector $\mathbf{w}$.
2. Let $\mathbf{w}+ = E\{\mathbf{x}g(\mathbf{w}^T\mathbf{x})\}-E\{g'(\mathbf{w}^T\mathbf{x})\}\mathbf{w}$
3. Let $\mathbf{w} = \mathbf{w}+/|\mathbf{w}+|$
4. If not converged, go back to 2.

DEPT OF ECE

The expression in step 2 can be derived using Kuhn-Tucker conditions and Newton's updating algorithms.

In practice, the expectations in ICA algorithm must be replaced by their estimates. The natural estimates are of course the corresponding sample means. Ideally, all the data available should be used, but this is often not a good idea because the computations may become too demanding. Then the averages can be estimated using a smaller sample, whose size may have a considerable effect on the accuracy of the final estimates. The sample points should be chosen separately at every iteration. If the convergence is not satisfactory, one may then increase the sample size.

To prevent differint vectors from converging to the same maxima we have to decorrelate the outputs $\mathbf{w}^T\mathbf{x}$. We use Gram – Schmidt orthogonalization to achieve it. The below steps are used after very estimation.

1. Let $\mathbf{w}p+1 = \mathbf{w}p+1-\sum\mathbf{w}^T_{p+1}\mathbf{w}j\mathbf{w}j$
2. Let $\mathbf{w}p+1 = \mathbf{w}p+1/(\mathbf{w}^T p+1\ \mathbf{w}p+1)$

<u>5. Frequency domain approach.</u>

The mapping from s_i to x_i is assumed to be a linear model A, as from eq (5)

$$x = A\,s$$

This equation describes an mixture model. Unfortunately this model is rather incomplete in the case of sources recorded in a real acoustic room environment, as sources are convolutively mixed. This problem becomes

$$x = A * s$$

where * denotes convolution. The purpose of ICA is to find an inverse model W of A such that,

DEPT OF ECE

y =W* x

and the components of **y** becomes as independent as possible. These mixtures are referred to as convolutive mixtures. This makes it then very computational expensive to work with in time domain. But if the equation is transformed into frequency domain the convolution transforms into a multiplication.

$$X_f = A_f \ S_f \ \ldots\ldots\ldots(14)$$

and

$$Y_f = H_f \ X_f \ldots\ldots\ldots(15)$$

This means that it is possible to apply the same negentropy method for frequency domain samples of **x**(real or imaginary part) subscript f representing frequency domain.

6. Acoustic Beam Formation

Array signal processing is a research topic that developed during the late 70s and 80s mainly for telecommunications, radar, sonar and seismic applications. The general array processing problem consists of obtaining and processing the information about a signal environment from the waveforms received at the sensor array (a known constellation of sensors). Commonly, the signal environment consists of a number of emitting sources plus noise.

Exploiting time difference information from the observed signals, one can estimate the number of sources present in the environment, plus the angles of their arrival towards the array sensor. The use of an array allows for a directional beam pattern. The beam pattern can be adapted to null out signals arriving from directions other than the specefed look direction. This technique is known as *spatial filtering* or *adaptive beamforming*.

One can set up an array of microphones and apply the techniques of *adaptive beamforming* in the same way as in telecommunications to perform several audio processing tasks. We can enhance the received amplitude of a desired sound source, while reducing the effects of the

DEPT OF ECE

interfering signals and reverberation. Moreover, we can estimate the direction or even the position of the sound sources in the near field present in the room (*source localisation*). Beam forming assumes some prior knowledge on the geometry of the array, i.e. the distance between the sensors and the way they are distributed in the auditory scene. Usually, linear arrays are used to simplify the computational complexity. In addition, optimally the array should contain more sensors than the sources in the auditory scene. Exploiting the information of the extra sensors using *subspace* methods, we can localise and separate the audio sources.

7. RESULTS

7.1 WHITENING AND PCA

The below figures represent the output PCA done using Eigen Vector Decomposition i.e diagonalization of matrix A. For the reasons mentioned in formulation this is not a good method of estimation.

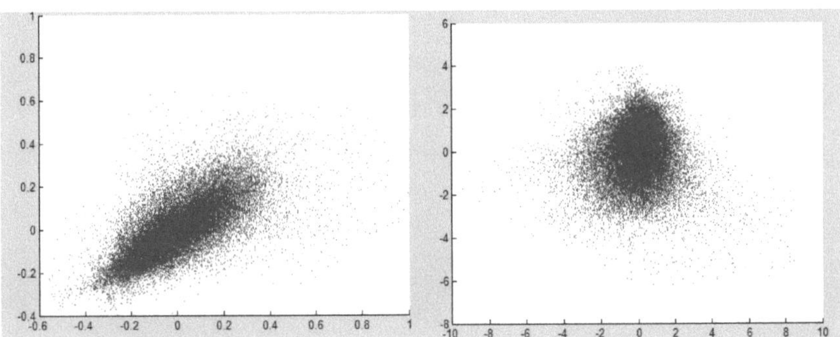

Figure 10 Whitening and PCA.

7.2 NEGENTROPY METHOD

The below figures represent the outputs of negentropy method for standard signals. For natural speech signals the estimation is not satisfactory as in practice we need to consider convolutive mixtures. Thus we go to frequency domain methods.

DEPT OF ECE

7.2.1 TIME DOMAIN METHOD

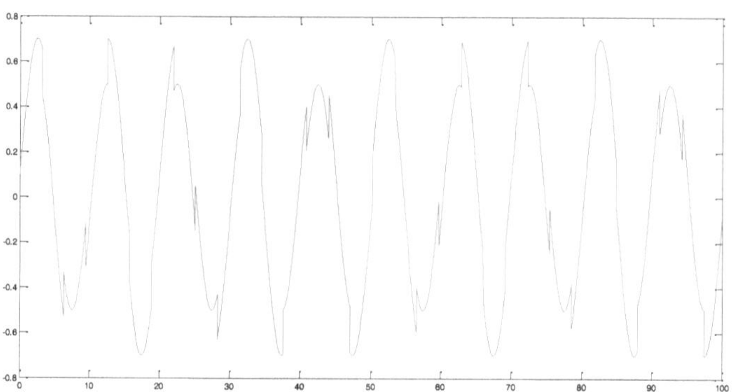

Figure 11 Mixed signal from source.

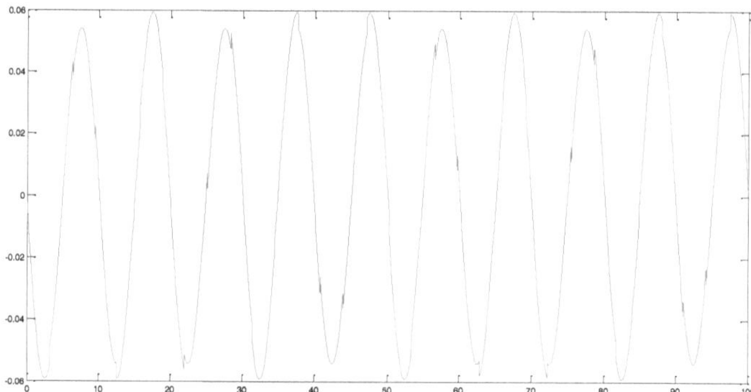

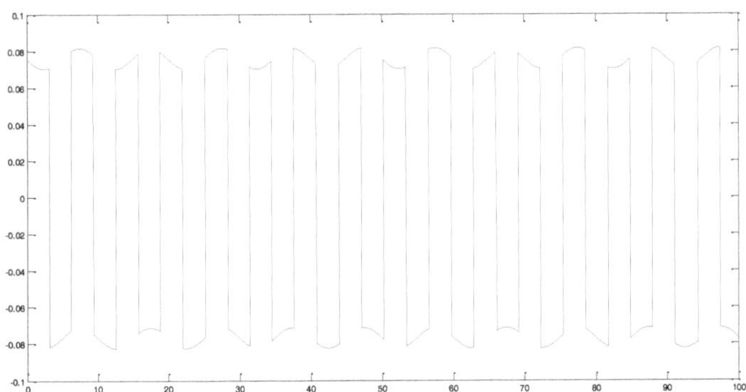

Figure 12 Separated signals

7.2.2 FREQUENCY DOMAIN METHOD

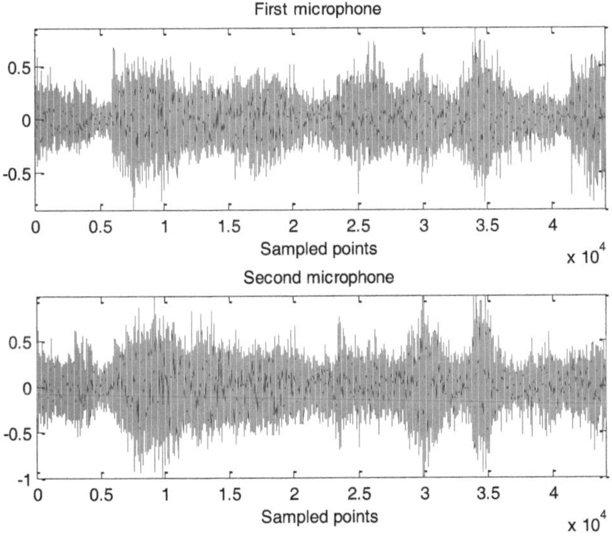

Figure 13 Mixed signals.

DEPT OF ECE

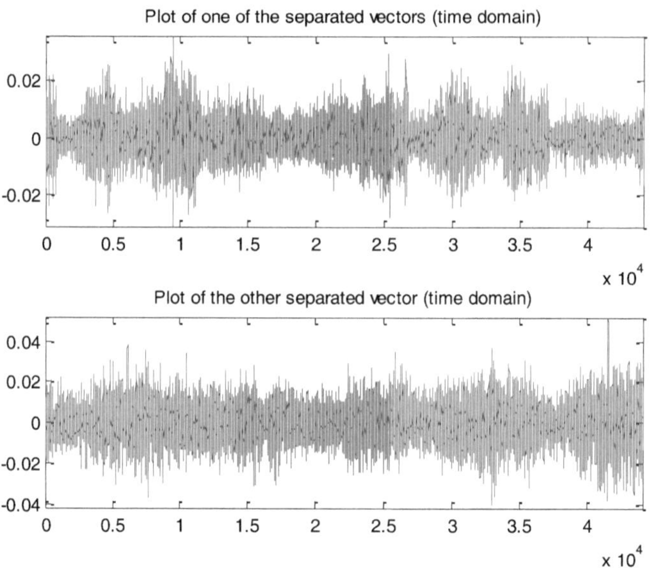

Figure 14 Separated signals

7.3 ACOUSTIC BEAM FORMATION

The below figure represents the beam formed in the direction of the source in MatLab
simulation.

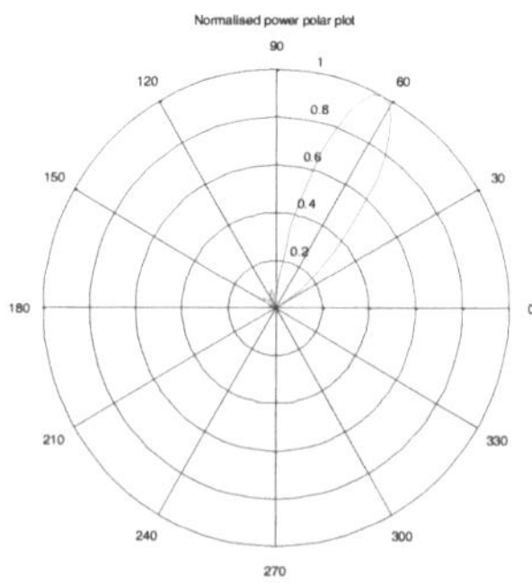

Figure 15 Beam formed in the direction of source.

DEPT OF ECE

8. REFERENCES

1. Independent Component Analysis: Algorithms and Applications
   Aapo Hyvärinen and Erkki Oja
   Neural Networks Research Centre Helsinki University of Technology.

2. Audio Source Separation using Independent Component Analysis
   Nikolaos Mitianoudis
   Department of Electronic Engineering,Queen Mary, University of London.

3. Real-Time Implementation of a Combined PCA-ICA Algorithm for Blind Source Separation
   Magnus Berg and
   Erik Bondesson